Praise for *Video Makes Cents*

"In the 20 years that I have been putting together video events for Fortune 500 companies, I have never seen better. Your creativity and professionalism are unsurpassed."

— Ray Hollinger, National Sales Director, IHS

"Your final product has been in the hands of our sales force for less than one week and they are reporting rave reviews from our customers. Thank you for a real masterpiece."

— Robert E. Delaney, Chairman,
Gamajet Cleaning Systems, Inc.

"You and your staff are artists and as such, have painted a wonderful, lasting impression of our company that will be instrumental in sending our message to our present and future customers."

— Andrew W. Barton, Jr., President,
Food Sales/Mar-Dru, Inc.

"We could not be more pleased with the final result. The tape conveys the impeccable professionalism that we strive for in our practice. That same professionalism was certainly well represented in the way VideoNet executed the production of our tape."

— Joseph A. Stigora, Ph.D., CCC-SLP,
Baxter Speech & Fluency Center

"Your professionalism was evident from the planning stage, through the actual shoot to a product we can point to with pride. Thanks again for a job well done."

—Richard S. Cummins, CFO,
Vantage Systems, Inc.

"From day one you knew exactly what message we wanted to convey to our customers and you perfectly captured this message..."

—Diane J. Gring. Public Relations Coordinator,
Chester County Library System

"I want to congratulate you on a job well done. The video you did for us at the time of our merger with Xerox was truly effective in communicating the intended messages to our valued employees."

—James A. Honigman, VP of Human Resources,
Xerox Connect

"After meeting with you and following the steps you had suggested, our product video turned out great with little effort on our part. Looking forward to working with you again."

— Joseph A. Valentino, President,
ARCM Corporation

"The videotape turned out better than expected thanks to your focus. I would like to thank you for the creativity and professionalism you showed. We look forward to working with you again very soon."

—Michael Savage, General Manager,
Philadelphia Tramrail Corporation

"Your professionalism, assistance and supplying us with timely scripts are much appreciated. It seems you enjoyed working with us as much as we enjoyed working with you."

—Barbara Hendel, Comm. Manager,
Exton Station Community Association

"I was personally impressed with your professionalism in designing, producing and editing the program, and also your persistence in ensuring that we created a superb product."

—Michael E. Herron, Communications Manager,
ERM, Incorporated

"Now our marketing department has an extremely effective sales tool for mailing out to potential clients. Thank you for your hard work and creative effort!"

—Robert Euler, President' King Limousine &
Transportation Service

"All of our Regional Directors and officers of the company were impressed, and said it was the best promotional video we've ever presented. Thank you for a job well done."

—Ken Peterson, CEO, LeTip International, Inc.

"The feedback from our clients has been fantastic, and I must admit your organization exceeded my expectations."

—Norbert J. Alicea, Vice President,
Human Management Services

"You and your staff were professional and always willing to express a creative idea when needed. Your writers, cameramen, make-up and audio technicians were also extraordinary."

—Jonelle Rodericks, Strategic Communications,
AstraZeneca Pharmaceuticals LP

"The video you edited for our WALK Awards Ceremony was a great addition to the program and started the evening on the right foot. As always, it's a pleasure to work with you."

—Amy J. Sauerwalt, Development Manager,
Juvenile Diabetes Foundation International

"I just wanted to let you know what a first-rate job you did on editing my tape for the Emmy's. In fact, it was so outstanding it garnered me an Emmy nomination for Individual Achievement, Best Reporter. I couldn't have done it without your skillful expertise."

—Janet Zappala, News Reporter, FOX Television

"The video was high quality, accurately depicted the hotel and most importantly, was completed ahead of schedule. Your attention to detail was greatly reflected in the final product."

—Mustafa L. Rashed, Account Executive,
Crawley, Haskins & Rodgers

"Your creative talents and technical efficiency enabled us to release our marketing video on time and within our budget. I will recommend you to other companies should the opportunity arise. Thanks for a job well done."

—Kevin Mulligan, President, CDK Mobile System

"I must admit that I was a bit skeptical that you would be able to get all the film that you would need by assembling different shots taken from only one meeting. The final product demonstrates how wrong I could be. We were all impressed by your ability to convey the benefits of the organization in a simplistic but comprehensive manner."

—W. Scott Lawhorne, President,
LeTip Chester County East

"I have never met a more dependable, honest and reliable person than Ron and I heartily recommend him to anyone who needs a man that can accomplish anything he sets out to do."

—George W. Brodt, Civil Engineer,
Department of Water & Power, City of LA

"We received our Shawnee 50th Anniversary video the other day and we are absolutely delighted. Your imagination and technique contributed greatly to a memorable video."

—Robert C. Fixter, J.W. Sparks Municipals

"Your ability to gain an insight into our business enabled you to produce and even better set of teaching videos for this year."

—John Schroeder, Vice President of Marketing,
The Sherwin-Williams Company

"Your interest, sensitivity to our needs and cooperation made the project a delightful one. We look forward to working with you again."

—Mansfield Bascom, Curator,
The Wharton Esherick Museum

"I want to thank you for the tremendous job you did assembling a quality tape that helps us define our mission to "Build Kids with Character."

—Ted van Beuren, Chairman, United Sports Foundation

"Ron's easy-going, professional manner made this whole project a delight for my company."

—Barry C. Lyngard, President/CEO, SensorPad Systems, Inc.

"...I would like to take this opportunity to express our sincere gratitude for the tremendous effort you put forth in producing our "Coatesville: Staying on Course" video."

—Paul G. Janssen, Jr., Former City Manager, City of Coatesville

VIDEO MAKES CENTS...

because it changes the way people think.*

* When done correctly.

by
Ron Strobel

Foreword by
Kirk Hanes

VideoNet, Inc.
Malvern, Pennsylvania

ISBN: 9780615369112_0-00

Published by VideoNet, Inc.
912 Saint Andrews Drive, Malvern, PA 19355

PRINTED IN THE UNITED STATES OF AMERICA

Library of Congress Catalog Number Pending

For all general information, contact info@videonetinc.com
www.videonetinc.com
610-647-3242

Cover Photo: © 2010 Steven E. Bayles ww.vtiphoto.com

Cover design by Jason Miller

Contents

Foreword

When Ron first contacted me about writing a Foreword to this book it caught me completely off guard. Then he told me why. "You know me as a vendor, client, friend and climber", he reminded me. Who better to invite a reader to take a closer look?"

Ron has worked with Disney, CNN, Discovery Channel, ESPN, and too many celebrities to count. He's President of VideoNet, Inc. a company sought after by the largest corporations in the world to tell their story. But his readers don't know what I know. They have no idea what this guy is made of, what he has done, what he can do. They don't know what I taught Ron, they don't know what Ron taught me. If they did they wouldn't be reading this Foreword, they'd be into the book by now.

Ron was introduced to me by a mutual colleague in the late 90's. I was a young Senior Vice President in a large corporation running a 500+ strong sales group. As a Colorado native my climbing background mapped well to growing a business in the sense you don't know what you are capable of until you try something with high risk and extreme consequences. I knew I needed to challenge these people to get out of old habits, try something risky with extreme consequences and succeed. In this case the reward was revenue growth.

Through a colleague's recommendation I hired Ron to tell the story that was in my mind at our National Sales Meeting. I needed something to grab their attention. This was a group of people with an average tenure in double digits. My story would be the first thing they experienced as it would open the meeting. The room was a very large with two huge screens on either side of the stage and enough sound equipment to boom. This would set the tone for the remainder of the week's events. It had to pop.

Our theme was 'The Competitive Edge'. The story would be based around summiting high peaks and the dangers therein. I explained my objectives to Ron and sent him on his way. When he returned to show me the first cut it wasn't what I expected.

What Ron produced as a non-climber at the time took my breath away. It began with a snow level shot looking up a couloir on a beautiful deep blue sky sunny day high in the Himalayan Mountains. Almost instantly an avalanche appeared rumbling loudly straight toward the camera then scooping it up and throwing it against rocks and trees within it's now fully white and tumbling lens view until the camera was destroyed. It continued with men in heavy winter climbing gear deep into an ice fall dropping an aluminum extension ladder across a 30 foot wide crevasse. The flimsy ladder clanked loudly in the quiet mountain when the top hit the other side of the crevasse and bounced to rest to form a foot bridge. The frame of the next shot was filled with a yellow winter climbing boot with metal crampons. The tenuousness of the climber was felt as his crampons slipped and scrapped on each aluminum rung as it tried to find purchase while crossing the icy abyss. It was mere seconds after I saw the head of the ice axe pounding in a single piton into ice that a camera shot from directly below looking up at the three climbers let me know they could easily die right there, right now. Then… one missed a foot hold and fell, straight over the camera and the accompanying music hit a heavy shrill. I jumped. The rope securing the climber whipped past the camera then jerked to tension as the climbers yelled. I was fully engaged and only a minute had passed.

Within the next few minutes it concluded with three climbers roped together and spaced far apart as they walked across the edge of an enormous snow ledge at an altitude where jets fly showing the grand expanse of mother earth far below. The music was orchestrated perfectly leading into the summit celebration at the highest point on earth followed by a slow fade to black in the dark room where I sat. I stared at the black screen. I didn't move and couldn't speak. My heart was pounding while my adrenalin said you just climbed Mount Everest… and survived. A voice in the room grabbed my attention but I didn't know who it was. "What do you think?" It was Ron, the only other person in the room. I paused, "I think, I think it's perfect, that's what I think!" I reacted, "How did you get all those shots? How did you know what I wanted?" He looked me straight in the eye, a slight smile crept to his face and he replied, "That's what I do". I remember it like it was yesterday.

Needless to say the National Sales Meeting achieved what I needed it to do. In fact the group asked for the video to be played again at the closing ceremony as "fuel for the road" in this new adventure called the internet that we were taking on.

Several weeks later I received a letter from Ron. He praised me for my leadership skills he had observed during the sales meeting and stated other gracious overtones then closed his document with a specific request, would I guide him up one of Colorado's famous peaks? I've taken plenty of young adults into bad situations why not an out of shape 50 year old man from Pennsylvania who I hardly know. "When would you like to go" I queried.

I picked Ron up at the Denver airport that summer. The next morning we hit the trailhead at

5:00AM and started our ascent. To stay on point I will forego many of the details that took Ron to a point in that day that changed his life forever. Yes, Ron followed my lead that day, tried something very risky with potentially extreme consequences but hoped to reap the rewards in the form of his first high summit. No, he didn't fall and obviously he didn't die but he did remind me why I don't guide anymore. Two thirds of the way to the summit Ron began experiencing hypoxia, a condition where not enough oxygen is being carried and absorbed by the body or brain to function or think normally. He was in tough shape and getting worse. Despite my knowledge of what was happening to him physically I could not get him to turn around. He kept putting one foot in front of the other, resting, and then doing it again. It is my longest summit on record but we made it. Although he was somewhat in coherent during our brief stay at the summit he is known today by the name I gave him at the top of that peak, Sherpa Ron.

After that trip he started walking everyday with a backpack filled with rocks around Malvern and lost 20 pounds. He would call me from his mobile phone, half out of breath and exclaim "I'm humping hills!" Every summer since then he plans a trip somewhere in the world, invites me, packs his camera and goes trekking around some of the highest peaks in the world such as Everest, The Matterhorn, and The Eiger. His goal is not necessarily to summit, his goal is to do something difficult, undefined and out of the ordinary that pushes his limits.

Why is this important to video production? Because it's what I know about the man's inner being, what's behind the project, which's behind the lens and what drives the end result. He already proved to me the mastery of his craft was far superior to others in his trade. He has an ongoing inner drive to always push himself in everything he does to new limits, including his work. I'd like to say Ron has pushed through all of the other video projects I have presented to him over time, but he hasn't. Ron will be the first one to stop a project if he feels the end result is not obtainable. For that, I am thankful.

Most videos don't make dollars or cents. I have an HD camcorder, my daughter has a Flip and there are a lot of people who call themselves professional videographers. Anyone can point, focus, shoot and call it a video. Ron knows that unless the person behind the production truly understands the way to elicit strong emotion around the content and deliver it in a believable way the message falls flat. A great corporate video must deliver a call to action or it doesn't make cents.

From the first video I hired Ron to produce, and every video since then, he has always shown me that he goes to extraordinary lengths to understand what needs to happen. The mountain climbing video is the perfect example. He had never climbed before but put together a message so compelling and so real it gave everyone thought for pause. And he was so intrigued by his pre-production research; he desired to experience it himself. It didn't matter if it was for business or for pleasure.

All can gain from his sharing decades of experience and wisdom in the form of this book. He's proved it for years in compelling productions for corporations, families and broadcast media networks. I strongly believe the reader will leave a better Marketing or Communications Manager, and stronger professional for reading it. Ron and I have obviously become good friends over time but I still attest to his skills and mastery of craft after exposure to others in his industry.

Sherpa Ron is one of a kind!

Kirk Hanes

INTRODUCTION

No matter how you are going to use video you will find this book to be a good resource. I will introduce you to my own personal business advice and a new tool. A tool that has been passed down through the generations first as colored drawings on cave walls, then in scrolls, then in language, then in print, then in film, now in digits. I refer to it as *Corporate Filmmaking*. You call it video, and it can move people to tears when done correctly. So welcome to the dynamics of persuasion, which equals more profit (cents).

This book is about discovering the power of this medium, coupled with a hundred or so business tips and techniques, to make a point or to sell anything because video can change the way people think. You must mess with their senses. It is based on my experience as a son, father, grandfather, artist, Art Director, CEO, freelance video cameraman, filmmaker, Producer, Director, editor, salesman, world traveler, and, yes, author. I have also been an employee and employer, which is important because my thinking considers both. What I am not is a *video guy*!

I am the founder of VideoNet, Incorporated. Our company works with owners of small companies and Marketing Directors of mid-sized firms to discover if they can close more business faster by simply bringing video into some aspect of their sales process. My unique ability is understanding the power and limitations of the filmmaking world and bringing that knowledge into the lives of clients, friends, and ordinary people. Video can generate revenue by distinguishing you from your competition in a sophisticated way, through a heightened conversational mode. We have perfected an extraordinary, educational story-telling format that is emotionally exciting and memorable. In short, I am not a Video Guy but an *Audience Reaction Expert.*

Consider for a moment that people don't seek out anything 'average' today. They look for remarkable and don't spend until they find it. In business a great video will enormously strengthen your competitive advantage. It will make it easier for your centers of influence to refer you to others and accentuate your product or service by clarifying *your unique wisdom.*

It's the most powerful, most effective communications tool in business today. It gives people a safe look into your business providing a higher degree of comfort. You can cleverly bring the large world out there into your small world through carefully constructed logic. You can also have third party endorsements in your video. It can provide a reason for a new encounter with a prospect or a client. Video is the most compelling way to tell your story by capturing the imagination of your audience. The more emotion, the more cents, because when done correctly, it can dramatically increase sales revenue. How? By entertaining and informing a population that has grown up with a preference for being shown and told rather than just reading. Hearing and *experiencing* rather than reading boring text.

TASTING CHOCOLATE SODAS

When I was a small boy, perhaps four or five years old, it was a real treat to stay over at my grandmother's house. It was so special to me because 'Millie' would always take me to the soda fountain at the local drugstore where I would get the same thing every time – a chocolate soda. And I could have as many of them as I wanted. But it's not the chocolate sodas I remember most; it was the door to this drugstore. It was a very weathered blue with chipping paint and three door handles. Not door knobs, handles. One was up high at an adult's eye level. One was where most door handles are, and one was very low, where I could actually let myself in – and that made it a very special place for me. More about this later.

Doing business in today's world means dealing with what I call the *ruthless rules of reality*. We're constantly bombarded with an avalanche of information through advertising, billboards, voicemail, television, radio, and of course, the world wide web. It's overwhelming and it understandably makes our prospects and clients more cynical than ever. After all, Nike tells you: *"Just do it!"* Do what? Snapple says: *"Made from the best stuff on Earth."* Well that really clears things up. What about McDonald's who says: *"You deserve a break today."* A break from what? Healthy food? No wonder prospects don't automatically trust us. They're confused.

The truth is they see you and me as one of many that may or may not solve their problem. They don't believe we're very special at all, let alone remarkable. Why should they when they can Google our service and more than 1,000 providers pop up? So how do you get their attention? You add another door handle. You let your prospects in. The answer is video, or what we call, corporate filmmaking. We believe every individual or business out there should have a video. This will allow your prospects to experience your company and to hear your company's voice. It becomes your well-crafted, easily understood offer, presented in an entertaining way where you first dissect, then demonstrate your strengths to differentiate you from the herd.

I know what you're thinking. Your company is too ordinary. You have no compelling story. Yes, you do. You just don't see it. That's why you need us. You're too close to the action. You don't get what's extraordinary about what you do, and even if you did, you couldn't put it into an entertaining story. It's the old adage *you can't see the forest for the trees.*

When it was time to re-do my demo reel, comprised of years of footage of two hundred fifty different companies over 15 years, I chose the best clips, then I gave an editor complete control to put it together in a logical fashion. I was too close to it to do the job correctly. I needed an objective source. Seems a little crazy, but it's true. All you really need is a great video Producer. At VideoNet, we will uncover what's remarkable about your product, service or cause from your customer's perspective. We write a script, hire a crew, shoot it, edit it, add effects, narration, music, and deliver it to you with you having approved every step of the process. Now you will be noticed!

This completed film will drive home the message that you *are* the logical one to do business with, guaranteed. Now you can put it on your website, hand it out as a DVD, take it on customer calls as a thumb-drive, have it on your laptop hard drive, mail it out, or include it as a link in all of your e-mails. Once you have it you will wonder how you did without it. It's your new digital business card. People will respect you more and perhaps for the first time, they understand exactly what you do, how well you do it, and how you differ from the competition. Bottom line, it'll get you to a decision faster. Truth be told, your company is only as ordinary as you've allowed it to be. Invite your prospects to come into your store. Add another door handle. Let them taste the chocolate sodas.

I believe the truth is most people understand that video makes sense but don't know where to start. They think it's complicated and expensive. This book can help clear up the myths and misinformation out there. It is based on the hard-fought experience of a network of video professionals who have consistently developed effective, award-winning, affordable video applications for growth-oriented companies to help them sell, explain, or motivate any audience.

MARKETING APPROACHES WITH VIDEO

Package your Process

What's the number one headache of most people in business today? Not enough time, they say. But what they really mean is that things have gotten very, very complex. We've reached a ceiling of complexity. Nothing seems to be simple anymore. Change is not only *in the air*, it is the air we breathe.

Most people are convinced they have to adapt to win at this game of persuasion with some level of predictability. Face it, unless you're a huge global company it's increasingly hard to consistently win at the selling game for very long. You must be willing to become more strategic. With a strategy that you believe in will come a sense of calm to the complexity of constant change.

Continually try to upgrade your role or organization by operating in a more strategic fashion by putting predictable systems in place. Simplify processes into steps that can be measured and analyzed. Make complexity disappear for you, your business, and your life. An aspect of making your business simpler is conveying a more simple message, a defining statement that is laser-focused and easily understandable by your market.

At our company, VideoNet, Inc., this is our defining statement: *We work with owners of small companies and Marketing Directors of mid-size firms to help them discover how they can close more business faster by simply adding video to their sales process.*

It tells who our target market is (owners of small companies and Marketing Directors of mid-size firms), quickly communicates the problem (not enough business and it takes forever to close a sale), and offers a possible solution to their pain (add video to the process). It's simple.

You can do the same thing. Be absolutely clear about what service you provide, or what pain you solve, or what outcome a prospect can rely on should they want to hire you. With video you can package an idea that's easily understood and then grow your company by solving some challenge that your prospects continually have. Provide piercing insight so they can clearly see through their own complexity.

Message Overload

Today we're all bombarded with marketing messages and advertising to the point where we believe very little of it. I like to read the back of any shampoo bottle and I am amazed at what this product is going to do for your hair. It promises things that are truly unbelievable. It makes claims that are ridiculous. Go ahead. Try it and you'll see what I mean.

You and I operate in the real world. We cannot get away with these kind of claims. It's more crowded, mature, and competitive out there, so today you have to really earn someone's business. How do you do that? Well, video can remove the clutter on the landscape.

Buyers are looking for business experts and buying decisions are being made at higher and higher levels. Purchasing has become consensus oriented. You can speed up the decision process by putting your offer in a consolidated form - the so-called "video elevator pitch." Simply make a movie to match your message and for the first time grab your prospects by the eyes and ears.

Why should you consider this? Because change is here to stay. There's no way to avoid it, and clients want innovation. Selling is a mental game. People make decisions fast and justify them slowly. Your prospects hunt out mutually beneficial, empathetic relationships, and you can create that spark by appearing to operate way beyond your normal job description in your well-crafted video.

With regard to change, get used to it. This is the new normal. Wake up to that idea and stop fighting it! Surround yourself with people who are willing to take on this challenge of continual change. Use it, and you won't be abused by it. Network with others that you trust and refer business to them, because after all, selling is a productivity game.

Again, we operate in an over-communicated society. How do your prospects cope with an on-going avalanche of words and images? By a selection process. By focusing. They dial out everything else, so don't try to appeal to non-customers. It takes too much energy. Be prepared to give up certain segments of the market. Deal with reality.

Sacrifice is an important part of a corporate strategy. Define who you are and who you are not by *your target market* because your client will selectively choose what to pay attention to. How do you get them to pay attention? Be clear, yet profound. A single word can be more powerful than a mission statement. Think your message through and hope that the story you tell or demonstrate with your video can help you own a piece of your customer's mind.

We all have what I call "belief-windows," and people today suffer from too many choices due to an over-abundance of everything including information. To your prospects there are plenty of you out there.

All they care about is value creation, so bring your specialty into their future by creating value in their world.

Remember, your audience, aka customers and prospects, are not you. Selling is selling yourself. Discover your prospect's vision of the future then build on that relationship, that dream, through your relationship with them. To do that you need to become a trusted advisor. You need to clarify the confusion by quickly changing beliefs they have about you and your products. It's very smart to identify your target market's pains, and then communicate the products or services you provide to make those pains disappear. This is an example of a simple strategy.

Rethink and recement your message to cut through all the advertising and marketing noise out there. Develop a quantifiable approach instead of a hope and a prayer. Consider changing your shotgun approach. Think rifle. Think arrow. Think laser focus. Measure what happens then either abandon your plan or modify it. Readjust, then measure again.

Today, more than ever before the customer *really is* in the driver's seat. Uncover a common frustration then be truly remarkable at solving it. Otherwise you're invisible – one of many. Claim a simple idea as yours. Constantly refine, test, support, and position your talents. Your offering must be innovative, indispensible and inspirational. In a video you can animate your value proposition, so be as different as possible in at least one important way. Shape and sharpen your message to ensure staying power. Create that tipping point because boring leads to failure. Boring gets you a "no." Your marketing *is* your product or service, so bring video into the equation.

Expect a Snap Judgment

Today, a snap judgment is probably all you'll ever receive from a prospect. At least it benefits you to believe that. Why? Because if you work on that premise you will clarify a simple message and focus only on that. Great advertising is seduction. Video can seduce with language, music and images, and back it up with real evidence, which is your track record of achievement. Confusion is the real enemy here. Claim a simple idea as yours then position your company with clarity because if you don't, prospects will leave with nothing more than a collective hunch as to what you do. What is your key claim? Give one concrete fact that supports it. Marketing is warfare, not word-fare. Differentiate wisely and lock in customer loyalty for life. Find the 'pixie dust' that makes your unique pieces fit together and then develop a constant, unwavering focus on your customer.

People don't respond to boring, and when they don't respond, they don't buy. So differentiate or die. Nothing penetrates resistance better than a dynamic story, and video is a story telling medium. Describe the story that makes your pieces fit together to reveal the entire puzzle, which is the solution to your customer's pain. That's a rational way to win more business.

Traditional selling will continue to die a slow, much deserved death. Today people will pay for exceptional insight, so find out what your prospect has already decided and why, then feed into that. Find out what outcomes your customers want then invent a delivery system for that actual product or service. Buyers don't buy what they don't understand. Build on existing ingredients they already have in their heads and then create new combinations. Customer satisfaction is worthless. A great video can promote customer loyalty which is priceless.

Becoming indispensable means being the 'default choice' of your customers. It's what we all want; so develop a constant, unwavering focus on the customer by organizing and packaging your story. Communicate a simple idea with enormous ramifications. That will create momentum. You need to let go of what used to work to make room for what will work next.

Thankfully, the brain is wired to ignore or filter out information it has heard before and already knows, so make your core message different. That's why in our world we're not video production experts, we're *audience reaction* experts. Would you rather have a prospect who is mildly interested after seeing your video, or the prospect who buys after seeing your video because every question he or she had about your company and the purchase have been answered? Case closed. Reveal the *secret sauce*, and they may not buy, but they will respect you as someone different. Mixed messages are the kiss of death. A confused brain does not buy, so when you find a unique way to communicate your niche, exhaust it.

Never tell someone something that you can show them. Structure reduces complexity and creates accountability. Demonstrate your structure and your customer will see you more as an advocate who legitimately puts their interests first. Forget the money part of the equation. I know that's hard to do, but it's worth it. Play like you owe this person a giant favor and your intent is to overpay them back. The money will come. Don't sell them so much as help them.

Here's another example of the power of video. Where is your company, or as I like to say, "Where does your company live?" It may seem like an odd question, but how would you answer that question?

The largest company in America is currently ExxonMobil. Where does ExxonMobil live? You might say at 5959 Las Colinas Boulevard in Irving, Texas. So if you were to go there and stand in front of the building where would you point? The building itself? Is that the company? Is it the sign ExxonMobil? No. You go inside the building. Is the receptionist ExxonMobil? No. You ask to see the CEO. Is the Chief Executive Officer of ExxonMobil the company? No, not really because he's got to answer to a board and they've got to answer to the stockholders. So, where is ExxonMobil? Or, let's wind things back to your company. Where does your company live? It doesn't live at your address. It doesn't live inside your business card. It doesn't live in your chair that you sit in every day, nor does it live on your desk or in your computer. I say, your company lives inside the minds of your clients and prospects. Let me explain.

Most of the world has never heard of, nor will they ever hear of your company. But for the very small fraction of people who have heard of your company, that's where your company lives. My company, VideoNet, lives in the minds of about 250 current clients and some of their employees, my family and friends, a few thousands prospects – and now you. There's nowhere else that it lives. Their thoughts about VideoNet give it life. How they feel about my company is designed to be what I've been able to project. So one can say that my company lives differently in the mind of each prospect and client, and it lives for them how they've shaped it, and how they understand it based on their relationship with me. How they feel is determined by this book, our website, what they've heard about us, our employees, what they've seen on video, etc. That's very important. Why? Because a properly crafted video can change the way people feel about your company. In fact, I can assure you that it will change how they feel. What I cannot tell you is whether it will change for the positive or the negative, because that is entirely up to the effectiveness of your production.

In my personal demo reel (on our website at www.vidoenetinc.com), this four minute video is designed to first reveal my experience as a freelance cameraman for ESPN, CNN, Disney and others. By mentioning the names of these successful companies, I hope to be put in the same league as them. Then I allow my audience to see me working on an industrial video shoot and then in an interview situation. This allows my audience to see me working and that I am in control of the set and appear to know what I am doing with actors and equipment.

I introduce them to the concept that I am fun to work with and involved in every aspect of the production. Then I show my audience what I do with my free time, which is climb mountains around the world. This is designed to allow my audience to get the concept that every video project we take on is *"similar to climbing a serious mountain."* Both require planning, endurance and professional execution. This comes through at the end where I am shown crossing a dangerous foot bridge on my way to Mount Everest in Nepal. A strong finish to a simple production designed to express my confidence and competence at anything I decide to take on.

Prospects come away feeling they know me and firmly believe that I can produce a good show, which was the complete objective of the entire production. Remember - inform, impress, and influence your audience, to change the way your company lives for them.

Your Front Stage and Back Stage

Here's a question. When you go to see a good play do you find yourself captivated by the action on the stage of the cast, the action, the lighting, and the sounds? Notice that you're only allowed to see the front stage of the performance. You know that there's action going on backstage, but you never get to see that - only the people who work in the play get to see all of that action happening back there. You know it's going on, but it's invisible *on purpose.* If prior to the performance you were allowed to see, appreciate and experience all that's happening back there at a frantic pace, you would be even more impressed with the actual performance on the stage. Yes?

Well, understand that every business has a front stage and a backstage. Video can show your clients and prospects exactly what you want them to see of your backstage operations. You can show them only what you want them to see; only what you're proud of with regard to your backstage operations. Focus on demonstrating your unique yet unseen process, and for the first time they might clearly understand exactly what it is you do, how well you do it, and most importantly, how you differ from your competition.

A good video Producer will take advantage of your back stage and make everything perfectly understandable. Your story is definitely worth telling and this will help convince your audience that you are the logical company to do business with, and if not, at least it gets you to a decision faster.

Time for an Experiment

What's your favorite movie of all time? Think about it. We all have a favorite or favorites. What about that movie do you most remember? You saw it so long ago yet is still has staying power - the action, the explosions, the music, the drama and the kiss. You remember it because it touched something inside of you that made a connection with your emotions and it has stuck with you all these years. All great film Directors know how to do this. A great video Producer or Director should know how to push the right buttons to inspire your audience (your clients and prospects) to take the action you desire. But not everyone can produce a predictable reaction, so when you find a great video Producer it's video gold!

After all, America has always had a love affair with the film world. Why not take advantage of that in the business world? After all we've already talked about the notion that your prospects are prewired today to regard any proposal that you make with suspicion and cynicism? Use the film techniques available to you to create a proven formula to communicate your cause, offer, product, or service that will inspire people to action. At VideoNet we have eight trademarked processes or formulas that our clients get to take advantage of (see them at www.videonetinc.com).

The Haunting Question

It seems to me that most of us small business owners always have the same question rolling around inside of our head: *"Where is our next customer coming from?"* Isn't this the question that most of us start our day with? After all, the purpose of business is to create customers, right? So the real question should be, "How do I build a predictable pipeline of work to support the cash flow my family needs?"

Someone asked me once, *"What are you building?"* That question really threw me because I never thought of my small business as building something. It took me quite a while to answer, but I said, "I'm building an income stream." Maybe you are too. To improve that income stream in your particular business, start by understanding that people never actually buy products and services. They buy outcomes.

Think about it. I say people don't need drills, and yet there are hundreds of thousands of drills sold in this country every year. Nobody needs a drill. What people need are holes and that's why they buy drills. To become a trusted advisor to a prospect, develop a sense of urgency because people buy on emotion. Listen carefully to their problem, and when you think they get that you understand their problem inside and out, offer them your solution on video. Video is a duplication tool that can deliver your unique offer over and over until someone reacts. It could be just that simple.

Extending Your Reach

To get the word out about your offer you may need help. Try partnering with associations and industrial groups, but first ask some tough questions. Do I join this group? Do I join that club? Do I go to this networking meeting? Do I offer to speak to that organization? Here in the 21st century there are new demands on us as business people. And there's a new science of selling and persuasion now called social networking.

Making new friends is wonderful and very valuable, but that's not what this book is about. This book is about profit. Partner with others that will bring the *ripple effect* into your business, by duplicating your efforts to increase profits. So first decide if the members of the group you are considering joining are your target market or have access to *your target market.* If not, don't join for new business because it's not there. If they meet that criteria then join, and do what you have to *'make friends and influence people'.*

Get your network and all your prospects to your website, and once there, have a no-risk offer to capture their e-mail addresses. Once you have their e-mail address educate them about how what you do relates to them and how you can solve a continuing problem that they have. Build trust because today trust is more essential than ever to the sale.

Keep in mind that everyone is not a prospect so don't chase everything. Take a close look at your last five or six big deals with great clients. What common problems did you solve for them? These problems will describe your target market, your real prospects. They are your best prediction of who might want to do business with you in the future because of your relationship with them in the past. Discover the attributes they all have in common like geographic location, gross sales, employee count, and executive contact level. Now where can you go to get in front of these people or to simply communicate with them? This is simple, yet profound, right? Consultive selling always looks and sounds like common sense. Find your core message and make it interesting by giving your marketing efforts a common visual language. Educate your prospects and clients about your wisdom through an entertaining, effective video.

Have the confidence not to second-guess your self. Offer clear-cut, remarkable solutions and always be prepared to give up a segment of the marketplace. Deal with the reality of your own resources.

Becoming Remarkable ASAP

Vision is the starting point for every creation. How do you envision your year going? Will you make your goals? Do you have goals? Are they written down? Are they measureable? It's very hard to compete today without a specific measureable game plan through tactics you have established. It has been said, "What gets measured gets done." Knowing that, have the wisdom to generate and visualize what you want, then, work backwards to devise a plan with reasonable expectations and timelines. Here's the secret sauce: All outcomes are feedback – the good, the bad, and the ugly. Use the results, or lack thereof, as feedback to adjust your plan, or in today's terminology, 'de-bug'.

Every owner of a small business needs to understand and believe in a strategy before any tactics can be deployed. One such strategy is to take on the position that your company is remarkable. Once you believe that, then you have to convince the rest of the world. How do you get them to see your company as remarkable? How do you sell that idea? Painstakingly!

Come up with items that are unique to you and you alone and give them a name. At VideoNet some of our unique items would include 'Our Rocky Balboa Narrative Arc', 'Our Risk-Reduction-Preproduction Questionnaire', and our '200% Money Back Guarantee'. It's these items that make help make us remarkable, and yes we market them! Give your clients and prospects the ammunition they need to rationalize their decision to buy from you. Continue to provide effective, long-term, beneficial solutions with superior service. Now *that's* remarkable.

We're all in a sense, a product. Can you explain to me in five minutes why I should do business with you? Can you do it in one paragraph? One sentence? One word? Try it out. Break your strong points down into thier essence. Clearly describe your specialty and communicate it effectively. Make one key claim then give one concrete fact that supports it. This will allow prospects to get a firm grasp on what is obvious to you because not completely understanding your offer can be highly detrimental to closing a sale. At VideoNet, we do it with a four-word promise, *A Clear Competitive Advantage.*

When a corporate video is done correctly it can have an enormous impact on a company's bottom line. Why? Because all of us underestimate the influence of television in our lives. Right now I am dictating into a recorder with the television on quietly in the background. Television changes how we think constantly, need we even debate that? *Video is television*, and it can change the way people think. At VideoNet our only focus, day in and day out, is to ultimately produce an audience reaction that leads to a true return on our client's investment. And if we do that, they hire us again.

How we consistently accomplish that is because we start by providing an intelligent assessment of how you currently use this powerful media, and perhaps, more importantly, how your competitors are using it. If you're a Fortune 1000 or a growing small business and you're looking for a fundamental strategy and process that will enable you to be more productive and profitable, your search is over. We are the masters at value creation through turnkey corporate filmmaking services that can help people achieve major on-going marketing and communications goals.

Once your video is shot, we say, *the magic happens in the editing*. We have a full editing studio where clients can observe and contribute to the creative process. Our state–of-the-art Final Cut Pro system uses the latest technology to deliver stunning, standard definition and high definition productions through the intelligent use of various custom software applications.

We have access to stock footage photos and video clips and a complete sound effects library. We have professional voiceover talent and the fastest, most experienced, congenial editors on the planet who understand the three act storytelling process and have knowledge of most digital movie making tools available today. It is through the use of shot positioning, narration, sound effects, music, and graphics that can make even a normally quite dull event, product, or service, exciting and entertaining. This is the fun part of the entire process. Sit back and enjoy it.

Video can dramatically affect the way your prospects think. Depending on your age, you may or may not remember what I am about to bring up. On September 26th, 1960, Americans watched the first televised presidential debate between Senators Richard Nixon and John Kennedy. During the end of the debate Senator Nixon began to sweat. Now we all sweat, especially under the harsh hot lights of the television world. And especially when millions of people are watching you and the presidency may be on the line. He was just being human. But what did the audience do? They decided that bead of moisture above a candidate's upper lip meant he was nervous, and therefore not telling the truth, and therefore not worthy of being President. Crazy but true that many believe the vote was decided on those beads of moisture. John Kennedy went on to win the election and the rest is history. People are a vessel of decisions made sometimes in a very strange way. Don't allow your prospects to decide your average.

Truth be told, your company is as ordinary as you've allowed it to be. To build trust, build consistency. Find your specialty and communicate it convincingly and consistently on video. Pull the trigger.

Educate, Educate, Educate with Video

You're the expert regarding your product or service, right? But instead of telling prospects how wonderful you are, create video content that is educational for them. It's *the cheese* in today's business world. Your goal is to inform, not to sell, and clients will appreciate your efforts. You've helped them, and they'll remember you. They will see you as an expert and seek you out to do business in the future. The information is inside your head, so just get it out there for prospects to use.

What's your story, and who's telling it? I firmly believe every business has a unique story, a history, a line of products, services, awards, and personnel. Usually prospects don't come unless you connect the dots for them by demonstrating how your expertise can resolve their ongoing problem. Here's where the art of storytelling really comes into practice. You can break your entire process into manageable pieces in a logical way that drives home the idea that will dramatically influence your audience to a quicker decision. Quite simply, you create a movie to match *your message.*

Napoleon Hill once said, *"To accumulate wealth, articulate specialized knowledge."* Become an expert in a particular aspect of an industry because power comes from organized effort. When your strategy, with measurable tactics seems to be failing, simply replace it with another tactic. Oddly enough, success is usually the culmination of controlling failure. Here's an idea...create an uncontested market space by pioneering a new market, your market. That's what we did. Instead of being in the video production market we're in the audience reaction market. It's a slight distinction, but one that puts us in a different league. It's like the 100-meter sprinter in the Olympics who crosses the finish line to win the gold medal because he beat the next guy by fourteen hundredths of a second. Is it fair? No. Is it important to notice? Yes.

Test your very own unique idea. Watch it break under the weight of evidence then rebuild it again. Lead with questions, not answers. Confronting brutal facts is part of life today, so focus on what to stop doing and you can produce a revolutionary leap in results. Take a percentage of your marketing budget and direct it towards video. Allow reason to triumph over reflex. Help create that moment in time when your prospect leans toward a "yes" to your offer. Switch from wandering generalities to meaningful specifics and watch sales grow. Allow prosperity to find you.

The good news is failure is all inside of your head. Make it small. Admit that your problems are a direct result of choices that you've made. Take reasonable risks because failing forward is actually a positive thing. It takes you out of your comfort zone and allows the possibility of true progress. Faith is merely taking the first step even when you don't see the whole staircase. Be on a voyage of discovery to attack and master your fears. Habits rule most of our actions and they're mostly all mental. What is my personal definition of a genius? Someone with an infinite capacity for hard work heading in a direction that shows promise. We learn practically nothing from a victory though it does feel great. Powerful information comes from a defeat.

Rocket from A to B, Not A to Z

We're all shooting for that ideal situation. That's how we're wired as human beings and as hard-working business people. But the ideal is never achievable. It's kind of like the horizon out there – you can see it, you can't touch it, but you can never get to it because by the time you get to it there's a new horizon. The best we can do is keep an ideal in mind as motivation and to be able to withstand the difficulties that are ahead and then to strive for mere progress, not perfection.

It seems to me that progress is not necessarily a function of discipline, but rather a function of having goal directed habits. Where you have a bad habit, find a new habit to override the bad habit. There is absolutely no need to market your product or service to people who don't really want your solution. You need to reallocate that energy and be marketing to the right people, *your target market.* Take the time to develop a common language with your prospects. It will give you increased confidence and promote clarity, which is what everyone wants.

The key to your success lies in being able to leverage you, or your company's, own unique talents so you can create value for your clients. Focus on what you're good at, what you do better than most, and what you enjoy doing. Through my experience most breakthroughs seem to come from negative experiences, and not necessarily the positive ones that we're all proud of. The key is not to avoid a negative experience, but to transform it into a positive one by using the energy it creates to initiate something new – some new idea, or process.

So let the notion in that *failure is good.* Do I prefer it over success? No. Never. But it does have some amazing qualities that most overlook. Failure can be a good thing if you wisely evaluate just what failed. This is called learning. In the computer world it's called 'de-bugging'. If you implement what you've learned from a bad experience into your system or strategy, it can provide the basis for a modified structure for future successes. Mistakes can just become tweaks.

Video allows you to tell your story in person, in a presentation, or to a group. You can put it on a DVD or a thumb-drive, on your website, or as a hyper-link in your e-mails. Give it a clear three act story structure so it cannot be misunderstood. Construct your core message as specific and provable. Third-party testimonials are great for this because they are an independent, third party source simply stating their opinions.

Your story is worth telling by you, and perhaps more powerfully, by others because videos can be easily sent as a referral. Become a recognized industry leader. Stand tall and stand ready to prove your assertions. Tell the honest truth and your prospect will appreciate that. Take them by the hand and lead them down a road, your road, to solving their problem. This is a vital strategy for future successful relationships.

Someone once told me, *"Where there's mystery, there's margin."* Your clients want effective, long-term, beneficial solutions and they want to be enticed. So give your video a mysterious name such as: "Myths about…"; "What you need to know…"; "You won't believe these tips regarding…" ; "Some great ideas to….!" Tap into your prospects desire for mystery and give them something good, something they can really use. They'll thank you by at least keeping your contact information if there is not a perceived immediate need.

Remember, trust is more essential than ever to a sale today. The marketplace is crowded, mature, and highly competitive. Buyers are looking for real experts providing measureable results. Plus, buying decisions are being made at higher and higher levels. This requires a new science of selling and persuasion because the old methods are sales suicide. The buyer today has a broad array of complex demands, most of which they will not communicate, so know your buyer through research and preparation.

Ron Strobel - 1999

CHOOSING THE RIGHT PERSON OR COMPANY TO MAKE YOUR VIDEO

It's time to produce your video. Who do you get to help? You want to remove as much risk as possible, so before choosing a video Producer we recommend you ask yourself some very smart questions before proceeding with any production. We call it our *Risk Reduction, Pre-Production Questionnaire*™, and it can be found on our website at www.videonetinc.com. It is designed to ensure success in telling your story, training employees, raising capital, demonstrating a process, launching a product or service, or whatever your ultimate goal. This document has been crafted through years of satisfying, demanding, hard-nosed, pencil-sharpening television Producers in Philadelphia and around the world. From this initial exercise, unique to us, you will benefit greatly, as have others before you.

Another resource we provide is the simple and valuable option of downloading our *9.5 Things You Need to Know Before Beginning Any Video*™ (see at www.videonetinc.com).

How many times have you tried explaining your product or service to a prospective client and have wished you could show them what you do? Video is a very powerful tool to convey your message consistently and professionally through demonstration. Many organizations use video for marketing and advertising and others find video a great educational and fundraising tool. Companies use video to train personnel, or to reach employees in remote locations with one-on-one messages where travel is not feasible or just too expensive. A great video crafts a message through a predictable narrative arc, the right talent, shooting, editing, graphics, and music.

If a picture is worth a thousand words, just think how much a professional produced video is worth to your organization. At VideoNet we're not Disney, or Twentieth Century Fox, or Paramount Pictures, or even E! Entertainment, though all of them have hired us on more than one occasion. We're just a small company with the ability to get you big results because we make a habit of getting to know your company from the inside out, before we show up with the crew and camera(s).

So get excited. Look at it this way. Welcome to the Academy of Purposeful Movie-Making. A great video will not only make your business stand out from the crowd, it will inform, instruct, and improve revenue. It can clarify your message and make it sparkle. It can force a faster decision and provide mass personalization, and as we say, grab your prospect by the eyes and ears.

Before hiring a video production company, ask to see finished shows similar to the one they will be creating for you. Ask to be able to contact those clients to discover if they were happy with their work. Know when the project is finished who has the rights to the show and who is going to keep the raw footage that the show was made from. Get any agreements made in writing and do not make a final payment until you are happy with the final result.

At VideoNet, we have a one page agreement that is quite simple and merely describes the responsibilities of each party.

We know our proven formulas will inspire people to action. A well-produced video creates permanent memories. It creates emotion, and people buy on emotion. Another way we like to say it: a video can make a great second impression. A well produced video can help your branding, create excitement, and cause someone to pay attention because they can watch it in the privacy of their own home or office. It can bring in humor and traction to help generate more revenue. Easier said than done, right? Not if you find the right Producer who will distinguish you from the competition by clarifying your products, services, or processes. This will educate your audience by communicating your experience and wisdom in an entertaining and exciting way.

If your website isn't performing as you wish, guess what? People get bored with just text. Video adds excitement, generates interest and traction and you rise higher in the search engine paradigm. And, have you heard the expression "death by PowerPoint?" Video can help that too. Bring a new level of sophistication to your PowerPoint presentations by including video in them, or replacing that glorified slide show all together with a true, motion picture like experience for your audience.

A great Producer will know you have a limited marketing budget, but video is no longer a luxury you can afford to be without. He or she will make it affordable and provide you with many uses for your new video some of which you probably can't even imagine at this point. A video is a great door opener that may just get you in to see the 'difficult to see' people. It can be watched at the convenience of your prospects on their schedules and in complete privacy.

How important is branding to you? Today it's the backbone of any business. Your prospects want you to offer a recognizable and distinct value. Your Producer should make sure your video will deliver a precise message that will cement your brand, provide clarity and distinguish you from the competition by systematically educating your customers and demonstrating your unique process. He or she can create a virtual tour enabling prospects to see your operation from the inside out and get a real sense of your core competencies.

Training videos lead the charge today because video trains equally, consistently and without error. Video offers you the ability to replicate a perfect training session for years to come and it doesn't eat, drink, or fly business class. It's quick, reliable, objective, and can easily be updated by reinforcing the correct way to do everything through simple, physical demonstrations that will make your workplace safer and reduce liability.

Take advantage of a video resource that specifies exactly what is okay and not okay. It can effectively communicate mandates, rules and regulations, and provide a provable record of who viewed it and when, for compliance. Simply put, video can demonstrate the right and wrong way to do things and explain the exact reasons in clear language by using real examples or animation.

Make sure you fully communicate with the Producer you've chosen that your production will be on time, on budget and on target, and with measurable results. At VideoNet, we provide an unheard of 200% money-back guarantee on any project we take on. After more than 700 video shoots from Philadelphia to Nepal, we've come to some conclusions. We know that a great video delivers a compelling message platform, yet less than 3% of all websites currently make use of videos. We understand you're apprehensive, but when done correctly, a professional video production will allow people to experience your company in a different way by engaging your viewers, banishing the doldrums, and connecting your dots. Complicated? Not at all. We've designed *Our Rocky Balboa Narrative Arc*™ that takes you from concept to delivered product in a headache-free, well paced manner. Once each section of the process is completed you receive a congratulatory V-mail™ from us indicating exactly what happens next.

Is it expensive? Projects usually range from $1,000-$6,000 per final running minute, but we'll bet once you have this new tool in your hands you'll wonder how you ever did without it.

How do you solve your problem with an ordinary video? First of all, your video won't be ordinary. It must be extraordinary! Every great filmmaker knows there is a proven, three-act formula, or 'arc', to achieving a Blockbuster hit (think of the first half of a typical Bell Curve). But sometimes, because film directors are extremely creative, they drift off course on purpose. They aren't concerned if their $200,000,000 extravaganza is a commercial success or not. They're pure artists.

At VideoNet, being in the predicable *audience reaction business* we cannot take that approach. We work with a time-tested formula, called *Our 8 Step, On Screen, Winning Formula*™ (see at www.videonetinc.com). It consistently moves people to action in an emotional way. They move to a decision faster, and it's usually a "yes" decision, because we've shown them the facts to justify their rationale. This is true artistic filmmaking that will continue to sell for years. One more thing: If we don't clearly see this potential in the beginning stage of our relationship we do not take on your project. Sorry - but we have learned that it is best not to take on a project that we sincerely believe is not going to be effective. This is a painful experience, but one we have learned to live with over the years.

How will a good Producer make your company exciting, and how do you know it will work? By a thorough understanding of your clear objectives, and then your Producer should get to know your company from the inside out. This process provides many ideas to clearly demonstrate, through strong visuals, that your company really is different and the one to do business with. This is the essence of good story telling. You'll know it's working when those who experience your story respond in a positive way. Why? Because for perhaps the first time, they truly understand what you're offering and how you're different from the competition. Create that tipping point conversation.

If you choose VideoNet to produce your video what can you rely on us to do? We will provide a fresh approach to mass personalization. Our objectivity and minimal knowledge of your business (at the beginning only) is actually a plus because we question everything. Our job is to peel away the layers that fog the brains of your prospects and clients, and to get to the compelling essence of your story and your specific offer. This becomes a new way to be seen and experienced in the marketplace. It's pure seduction. When done correctly a custom video has the ability to quite literally generate revenue, reduce costs, reinforce your brand, and reduce liability.

How can we put a value on something that doesn't exist yet? A I have said earlier, there is little clarity in the marketplace, only complexity. The promotional noise out there is deafening, yet nobody listens. So, how do you really get somebody's attention today?

You do it through a precise blend of common sense and emotion. All great film directors know this. We will capture the imagination of your prospect in a way that can't be ignored. We will laser-focus our knowledge of your company into a visual understanding of exactly what you do. In short, we change minds, and therein lies the value.

Your typical day is probably a combination of handling or delegating problems, a routine you've learned to live with. Why add another layer of complexity with a video? Because it's a new tool that you will learn to use just as you have a fax machine and cell phone. Thirty years ago they didn't exist. When we start out together, your input is always welcomed, but from beginning you decide how much you want to be involved. Our job is to keep you totally in the loop as we progress to completion through *our V-mail*™. Bottom line, we offer a script – to – screen, complete turnkey solution.

We provide video production services nationwide with jaw-dropping productions. Another reason why video is powerful is that anything that helps you prescreen and presort prospects is good. Nothing penetrates resistance like a dynamic story. Clients are *'experience dependent'* and you can provide the cognitive candy. Find a way to arouse their attention by producing a video that does just that. Learn to improvise from a collective database of information you have at your fingertips. You have it in your head. Just get it out onto the television or computer screen. Find a way to arouse, and then hold, someone's attention. All clients have a point of view. Change that point of view.

The essence of great marketing is putting yourself in a position to succeed. Some climb the ladder of success only to discover that the ladder is up against the wrong wall. Do you realize that most all of known human history can be rewritten in a simple little formula: Challenge, response. What are your daily challenges and, more importantly, how are you responding? Short-term satisfaction is like cotton candy. It doesn't last. Think long-term by finding the sweet spot in your field, the sliver of the world that you want to be connected to.

Your greatest asset may just be your database. Business is the world's greatest sport. Efficiency is doing things right and effectiveness is doing the right things. So, slow down and be deliberate. Test the waters by developing pilot projects because the future is always going to be fuzzy and vague. Learning is really adopting an attitude of discovery. When you choose the game, you inherit the rules. Form a practical blueprint for action. Be linked in with your clients' prosperity. We are!

THE PROCESS OF MAKING A VIDEO

Pre Production

Every video is different, but certain types of videos do fit a basic format or template. The basic procedures followed for every *marketing video* we produce are as follows: Once the client makes contact with us we introduce them to *Our Risk Reduction, Pre-Production Questionnaire*™. Usually after that there is a face-to-face meeting with myself or a qualified VideoNet associate; someone who is part of our team of video professionals. We would then research four or five of your competitors via their websites to discover how and if they are currently using video. Then we brainstorm potential ideas. This is usually done in an audio-recorded session with your sales staff. That audio recording is studied and eventually edited into what we call a *first draft script*. More about that in a moment. We have five decisions to wrestle with at this intial meeting:

1.) The key objectives of the video itself - What is it you want your audience to do or think once they've seen your video?

2.) The length of the video – Anything from 10 seconds to 2 hours (or longer).

3.) The uses for your video – Where will it end up? Will it end up on your website? Will it end up on a DVD to be handed out at a trade show? Will it end up on a thumb-drive, or as a link that you're going to send out in your e-mails? How will it be archived?

4.) The style of the show - This is very important because we can either be straightforward or be very clever and artsy with the video. A lot of times the objectives of the video will define the style of the show. We often refer people to our website (www.videonetinc.com), where there are more than 30 samples of different video styles from completed productions.

5.) The deadline or completion date – When do you need this video to be completed and a master of it in your hand for reproduction?

Then perhaps, comes one of the most difficult decisions – *the budget.* How much do you want to invest to accomplish your objectives? The budget, of course, is the backbone of any video and it will determine how clever, fancy and expensive we can make your production look. Obviously, the bigger the budget the more time and effort and talent that can go into your show, but it's important that you only have us produce a video that you can afford. Our normal payback time for any marketing video that we've produced, on average, is five months or less. Once we've met with your sales team, and have audio recorded that brainstorming session, we take that audio recording and any notes that we've made back to our office and produce a first draft script. Then we get that to you via e-mail and you tweak and improve upon it. Once we have an *approved script*, we create what we call a list of shots, that is, what videotaped scenes we have to shoot, and where.

It's at this time the camera format and screen size are selected. You might go with a simple small format camera, or a more complex, more expensive, large format broadcast type camera in either the 4 x 3 or the 16 x 9 formats.

Once the camera and equipment are chosen we move on to the selection of the crew. You should only have the crewmembers and equipment on your shoot that is absolutely necessary to complete the objectives of your video. Do you want actors in your show, or will you use your own employees, family and friends? Is there going to be an on-camera narrator, or are we going to rely on voiceover talent? Are we not having any narration whatsoever and simply relying on what we call talking heads, i.e. people to be interviewed?

Now we have a list of shots is created, the camera format and screen size selected, and the crewmembers and equipment are reserved. Actors and shoot locations are chosen and scheduled. Then we move from the pre-production phase, which is everything I've just described, into the production phase.

Production

The crew and any auxiliary equipment arrive on location or locations. Then the actors arrive and we handle wardrobe and any makeup issues. If we need releases signed by any of our actors or anyone who is going to be on the screen the time to handle that is now. Be sure to get releases signed by all employees participating in your show. The ramifications of not doing this should be obvious, especially if someone is fired, or hired by a competitor. Noise issues are handled because we want it to be as quiet as possible so that we get good, clean, crystal clear sound. Props and any other on-camera items are prepared and checked in and stored. Lighting is set.

Lighting is very important because video loves light. The more light you can produce on your actors, your product, machine, etc, the crisper it will look. The color will also be better, and everything will also to appear to be more in focus.

Now we actually do the videotaping itself where the director says "Action!" All our shots are captured, and then we also capture what is called room tone. It's the actual sound of the room, also called white noise. We do that because we may need it in the post-production phase in order to make the sound appear to be as clean as possible. The footage is labeled, fully secured, and transported to the editing facility. This is where we move into the third stage, which is called post-production, or what most people refer to as the 'editing phase'.

Post Production

Here we bring all the raw footage into the computer through a procedure called *digitizing*. That footage is color corrected and at this point we collect the client's electronic version of their company logo. We look at all the tapes and rate them via a VideoNet clip rating system selecting only the best ones. Then it's given to a senior editor who puts together the footage in a logical order, and then a master editor decides on the final arrangement and effects.

The narration is recorded and any animation, special effects, graphics and logos are added at this time. The music we use in a show is usually non-copyrighted because it is most cost effective. At this point I would take another look at the show before approving it being given to the polisher. A *polisher* is a master editor who comes in and looks at the show objectively and tweaks it to make it seamless. He makes the slight adjustments where we cut from one shot to another appear to be flawless and smooth. He also adds special effects with motion, color, layers and transparencies. The polisher makes our shows dance! That is the term I use when everything comes together to make a one-of-a-kind production exciting and memorable. The client doesn't see the show until after the polisher has worked his magic.

Multiple Uses

At this point we would produce either a DVD or put the show up on the internet for client approval and improvements. The client makes suggestions and then gives approval after we've made all the changes that they would like. At this point, it moves into the fourth and final phase of the production, which we call multiple uses. This is where the DVD digital master and duplicates are made along with some disc label art before being delivered to the client. We also make the production into various file formats for web placement. We can now cut the production up into separate little modules if you only want to use portions of your show for another use. Also, the production can be cut up and put into modules for any PowerPoint presentations you would like to dress up with some moving video. We always keep a DVD digital master just in case something happens to the client's and also so that in the future we can easily make tweaks and modifications, and usually at no charge for our loyal clients. That's the basics of our entire video process. Not our entire back stage, but some of it.

The Secret Sauce of VideoNet

We at VideoNet also have what we call *Our 8 Step, On Screen, Wining Formula*™. This is a formula that we've discovered captures and keeps people's attention so that your show gets watched. Our clients don't normally have major star celebrities, any explosions, dying scenes, or intimacy in their productions. Therefore, you need a proven template that works when all of that on-screen action and excitement is missing. The following is the template for a *marketing video* that we've found works best:

First, we fade up from black bringing in the client's logo with a carefully chosen musical score. People respond to movement so normally we try to move the logo in somehow by having it flip in, or come in from the side, or the top. With motion comes emotion.

The music sets the tone for the show and should be royalty free and, of course, not copyrighted. Now we reveal the title of the show. It's important that the title grabs the attention of your audience. It should be something very clever yet based on reality and we would suggest never more than one line.

Here is where we reveal the problem in a painful way that your prospect can relate to. Be dramatic. Be harsh, specific, and ruthless with the pain. In a world of marketing noise your video must get their attention. Your audience should actually experience the angst. We want to clearly communicate the pain that prospects wish to avoid. Then of course, we announce that our client's company can solve this problem because they are a pain solution provider. Be rational, specific, and confident. Validate your claim with supporting visual evidence showing them the problem disappearing. Now comes the part where you have to prove your claim with some irrefutable evidence. We suggest between three to five points that support your premise. Use unexpected implications in your core proof. Utilize a client testimonial for each of your points, if possible. If that's not possible, then use case studies. Simplify things by finding a universal language. Statistics are great to use here, but not exclusively. Distinguish yourself creatively while demonstrating competency and why you are the obvious choice to do business with.

Now, because people forget very quickly, recap the three to five claims that you've made. Tell them again in just a sentence or two. Connect the dots for your audience. Close your argument with an irrefutable summary. Here, we usually have some kind of call to action with a sense of urgency. After all, most audiences are hungry for direction and simplification, so tell them what their next step is. Make it very easy for them to find out more by having important contact information on the screen such as your company logo, phone, web address, and e-mail. The music builds up as the narration subsides and the music fades as the screen goes dark.

For a marketing DVD we recommend a length of no more than five to eight minutes. For a website video keep it to a length of 3 minutes or less. For a PowerPoint presentation go with a 30 second video module within your show and have as many of them as you'd like. For a tradeshow we recommend a looping, demonstration video only, with no sound.

What can a great video do for you? We've developed a list of attributes so that our clients may subjectively evaluate the performance of a past video project, or, develop a set of realistic goals for a future production. We call it *Our 23 Video Attributes Performance Evaluator*™ (see at our website at www.videonetinc.com). Video technology is fundamentally changing the rules of the selling game and it should ideally include some or all of the following benefits.

A great video should be an entertaining execution of your focused offer or story. It should speak directly to the decision maker delivering a compelling message platform. It should create differentiation that people readily grasp by dramatically connecting your dots. It should cleary demonstrate your core competencies and communicate your offer in a logical, repeatable, commonsense way.

Make everything understandable through simple demonstrations. Bring your entire world into your prospects' small world. An effective video can be great second impression, so make your company memorable instead of forgettable.

Make your company remarkable instead of invisible by allowing your prospects to see and hear directly from your actual clients through video testimonials. Allow your prospects to see your facility or your process. Take them through a demonstration or a virtual tour. Move people to take action or a defined next step by creating a tipping point conversation.

Build a case for a mutually beneficial, empathetic relationship. Clarify any confusion for your audience by providing a narrative arc, drawing ideas together and transforming information that you've provided into *knowledge they can use.* Be innovative, indispensible, and inspirational because great video is pure seduction.

Your video should have a shelf life of at least two years. Our experience indicates that people do not always make decisions logically. Instead, decisions get made quickly based on emotional connections. Great corporate filmmaking can significantly influence this incredible process.

The Six Myths about What does a Video Cost?

What does a video cost? Not a day goes by that we are not asked this question. To comprehend the answer you need to understand what a video comprises. After all, video is the best way to capture the complete attention of your client, customer, prospect, or employee. Listed below are the six myths regarding video (available as a PDF at www.videonetinc.com)

Myth 1: *The longer the video the more it costs.* This is not necessarily true. What you're really paying for is the time and experience of the Producer, camera person, and editor, and not necessarily the length of the video, which should be as long as it takes to meet your objectives. That could be 10 seconds, 90 minutes, or longer.

Myth 2: *A wedding guy is cheaper.* True, but you may get what you pay for. A wedding guy has his own gear, but he simply documents the wedding day. He or she is not necessarily a storyteller and that's exactly what you need. Documenting a wedding is different than selling a product or service. Get your money's worth by finding a freelancer with broadcast television experience who owns his or her own gear and can satisfy your objectives by telling a compelling story to make your case. References may be a very good idea here.

Myth 3: *I can do it as well in-house by myself.* Very unlikely, unless your objective is to shock your audience! Yes, amateur videos can thrive on YouTube, but they are mostly real life coincidences that happen to be recorded. They scare us, make us laugh, amaze us, but rarely sell us on a product, service or idea. For that, you need a professional to prepare a strategy and create an entertaining story that leads the viewer to a 'yes' or a crystal clear next step.

Myth 4: *It costs more for quality.* True, but the quality of what? The images? The soundtrack? The DVD itself? The process? The actors? The style? A great video is a composite of all these things. The Rodney King Beating Video was one of the worst quality videos I have ever seen. It had no color, was shot at night, hand held with bad sound. But it clearly met its objective - to demonstrate police brutality, producing a predictable reaction in people. This video and the subsequent police trial, lead to four days of rioting in Los Angeles that I got to experience first hand. Great images on-screen with no storyline or obvious action step are worthless no matter how beautifully executed. The real test of quality is whether or not your audience is being informed, inspired and influenced to a quicker decision. So if you mean the quality of the production to deliver on your expectations, I agree.

Myth 5: *All these different formats add to the cost.* Untrue. Once VideoNet completes your video we don't even charge you for providing it in other formats. The truth is if you have a good DVD master almost anyone can transfer it into any other digital format you desire. Digital is digital, just 1's and 0's.

Myth 6: *If I reveal my budget I will pay more than I had to.* Wrong. We suggest you provide a budget figure because this is how we like to work with our clients. Why? Because the client has an automatic guarantee that we will not go beyond their number. For example, with a budget of $10,000, we can currently provide a concept meeting - $500 – script writing - $1,500 – two days of video-taping at $1,450 per day, voiceover narration, about $500, three days of editing - $3600 total at $150 per hour, music and graphics - $1000 – DVD approval, master, including two hours of changes - $300 – master DVD and web files required will be made at no charge.

If you prefer on a la carte production professionals will charge a certain amount per day for videotaping which includes a two-man crew and all equipment with additional hourly charges for editing. The total cost depends on how organized you are and how complex of an edit you need. A good Producer should be able to give you a reliable estimate once you provide a script and possibly a video for him or her to evaluate. You should also explain at this time the style and quality you are expecting. Our videos usually range from $1,000 to $6,000 per running minute of a final production.

So what does a video cost? Well, what does a car cost? See the dilemma? Do you want a Kia or a Rolls Royce? Both can get you where you're going and we're here to help you with whatever you decide.

Video Camera Formats

Now for a brief discussion on video formats. We will cover standard definition video, high definition video, and film. Most of us don't think much about the images we look at on a TV, flat panel display, or on a laptop or desktop computer screen. That is, we don't think about what comprises the image because our eyes and brain merge all the elements of color, brightness and contrast into a picture we immediately see. But if you move up really close to an electronic image display and use a magnifying glass you can see the tiny, individual squares of red, green, and blue pixels (short for 'picture elements') that comprise the image. The smaller the pixels and the more of them that are distributed vertically and horizontally across the face of the screen, the greater the resolution, or detail, that we see in the image. The pixel count will also affect the cost of your production because a lower resolution camera costs less. The higher resolution costs more. Sometimes a lot more.

Standard definition is the equivalent of 480 interlaced horizontal lines stacked from top to bottom of the screen. The interlaced lines mean that the image, or frame, is made up of alternating fields of 240 lines that are scanned across the screen every 60th of a second. As the fields combine, or interlace, on the screen a full frame of TV in standard definition is presented. Anytime you shoot or have a video production and you're shooting with standard definition camera this is the image that you will get. This level of quality is fine for most corporate and web viewed projects.

If you're shooting with a high definition camera you get to take full advantage of the high definition ability to render spectacular clarity and detail that has a native resolution of either 1280 x 720 pixels, progressively scanned, with a wide-screen 16:9 aspect ratio, or, 1920 x 1080 pixels which represents a 16:9 widescreen image with 1920 pixels across each of the 1080 interlaced scan lines.

Progressive scanning produces a smoother more *film like* look, but the 1080i image actually contains greater detail. You can figure that the difference between a high definition production and a standard definition production will be about twice the cost. If you want to go all the way to a film production which is a lot more complex and complicated, requiring a different kind of crew, then you're talking two times the cost of a high-definition shoot. Usually film in today's world is not recommended unless you're making a high-budget feature film that millions of people will see on a large movie screen, or a commercial broadcasted nationwide on television.

We work with most all formats including DVD authoring and interactive shows and can deliver your final show on DVD or any web based compressed files you desire. Once approved, we will provide you with a digital master that will last forever. We also keep a master 'in house' allowing quick revisions or improvements for years to come.

WHAT MAKES VIDEONET UNIQUE

What makes VideoNet different? First look at our name. We are known as the place where you can have access to a time-tested *network* of video professionals. What does that actually mean? In the same way you would hire a contractor to custom build your home, you know that the contractor you called isn't going to be doing the foundation work, the framing work, the electrical, the plumbing, the roofing, the interior painting, and the stucco on the outside.

So, how does all that get done? He knows the experts to call, that's how. He surrounds himself with a 'unique ability' team of people and you pay only for the people that you need. VideoNet works the same way. We have relationships with the finest, most talented video and film personnel in the Philadelphia metro area who are able to perform any single camera or multiple camera production. From camera people, to sound technicians, to production assistants, to actors, to voiceover narration, to specialized equipment, like teleprompters and jibs and steady-cam apparatus. With VideoNet you only pay for the personnel and equipment needed to produce your particular production and therein lies the benefit.

In larger production companies, like most of our competitors, you would first talk to a receptionist who would assign your account to a salesperson. The salesperson would come out and meet with you to discuss your needs. Then that salesperson would take that information back to an estimator who would calculate the cost of the project. Then he hands it back to the salesperson who goes out and gets a 'yes' from you. Then once he gets that 'yes', he turns over the project to a Director who grabs a cameraman, who grabs a sound person and they go out to shoot your video. Then, they hand the raw footage to an editor. That editor puts together your project and then he hands that video back to the salesperson who comes back out to you and tries to get you to approve the project.

All of those people have to be paid. All of those people have vacation time and benefits. All of those people may reduce, or what I like to call, *dilute*, the original intention of the project. With VideoNet, it starts and ends with myself. There is no dilution, and you only pay for the people necessary on the project. They are all people who I can vouch for and who I have worked with many times in the past.

In short, *it's a subcontractor thing*. More of your dollars end up on the screen and less in health care and sick days for a multitude of employees. It's all about productivity. We bring the movie business into your business. We connect and create that spark, offering clear-cut, headache-free solutions.

My personal unique ability is understanding the power, and limitations, of the filmmaking world and bringing that knowledge into our client's worlds. Usually, my sole purpose is to increase revenue through the intelligent use of video to produce an audience reaction that will get prospects to a decision faster. That makes good business sense because both parties win.

What Promises Do We Make?

To every client that trusts us to provide them with some sort of a video solution, I make three promises that I usually don't make verbally but emotionally 1.) We will not take on any project that we don't believe has a chance of success. 2.) We will finish the project we start on time and on budget. 3.) I will never let my clients down.

One of my personal mentors is Steven Spielberg. I admire his work, as I guess most people in the world do. We all seem to agree that he understands how to make great films that you never forget. And herein lies the key to the motion picture business of corporate marketing with video.

I was fortunate to meet Steven in 1993 in Santa Monica, California. He talked about finishing his new film, *Jurassic Park.* It was sometime later that a good client of ours mentioned that the reason he outsources all of his video work to VideoNet is that to him, *"I was the Steven Spielberg of the corporate film making world."*

The rest is history. Our company has been very successful. The most important thing people seem to forget is that just because someone can record images in a camera does not necessarily mean that they can tell a real story. Someone has to have the expertise to arrange hundreds of images (or what we call clips) along with visual and sound effects, voiceover narration tracks, and musical perfection in such a way that your audience has a compelling, predictable reaction. This takes true talent and is remarkable when it happens. It is the distinction that separates the true professionals from the rest of the pack. We take pride in understanding the most basic rules of the blockbuster films of the past, and use these time-tested techniques to ultimately solve problems for our clients. Invest with us and your unique story will be brilliantly told to achieve remarkable results.

What differentiates VideoNet from our competitors? We have been noted as one of the Top 100 producers in America by *AV Multimedia Magazine*, the leading publication in our industry. We've been responsible for more than 500 broadcast quality television shoots. We've produced more than 200 corporate videos and have worked with many well-known celebrities.

We've won more than 15 national awards for excellence and have produced six festival films. We offer an unheard of 200% money-back guarantee and have provided video services for more than 250 clients worldwide. We have trademarked processes and handouts such as our Risk Reduction, Pre-Production Questionnaire, Our Rocky Balboa Miracle Arc™, Our 8 Step, On Screen, Winning Formula™, our Video in a Week Fast Track Formula™, and our V-mail™ communications process.

And finally, we have been approved and used on hundreds of occasions by national crew brokers, those in the business of hiring video crews for their clients, such as Assignment Desk, Crew Connection, Crews Control, Crews Star, Maslow Media Group and others.

MY PERSONAL HISTORY

People ask all the time how I got involved in this business, so I'll tell you the short and brief story. It starts way, way, way back. When I was a young boy, I think the first movie I ever saw was the *Wizard of Oz*. Those scary flying monkeys and green witch left quite an impression on me. Later on my family life wasn't the greatest during my teenage years. I never had a very good relationship with my father when I was growing up, and looking back, I think I went to the movies to escape. Escape what you ask? How about life, responsibilities, my dad, and authority of any kind. As an adult I can tell you people escape all the time. They escape in books and cell phones and computers today. They narrow their view and enter another world where they are more in control. So, as a kid, I would enter a dark, quiet environment and for about two hours I would enter another world with no responsibilities or concerns. A true pleasure for me, even in the 1960's!

Every Saturday my friends and I would walk a couple of miles to the Norris Theatre in Norristown, PA. A five minute black and white serial like Flash Gordon would show before the real film, and after the 90-minute feature movie was over we would walk back home.

To this very day I see a movie about once a week, and because of that, my clients get the benefit of the thousands of films I've seen. Some of which I've loved, some of which I didn't like that much, but all of which I learned from with regard to what we call beats, and lighting, and timing and what we call the 'arc of the film', and the three-act play.

Of course it didn't hurt that my father was an artist and that I went to college and majored in Art. Out of college I became a graphic designer for a publishing company, and then went on to become an Art Director for a television company. And then around 1980 I formed my own company to renovate small apartment buildings of six to 40 or so units. We bought the ugliest building on the street, remodeled it to look like the average building on the street, and then sold it as quickly as possible.

This is what got me involved with video because we would take video of the building *before and after* our renovation. To be able to buy more buildings faster we needed investors. An investor could see from our video what it looked like before (we paid $250,000 for it) and what it looked like after (we sold it for $500,000). It made a dramatic impression on them, and my company was very successful until 1994 when the California Northridge earthquake came rolling through Southern California and brought down most of our properties including damaging our home.

My wife, my two daughters, and I had a devastatingly negative experience of going through that earthquake. After that nobody wanted to live in Northridge, California, so I moved my family back to good old Pennsylvania and the suburbs of Philadelphia where both my wife and I had grown up. I took about a year off and decided what I really wanted to do to earn an income.

I had some money from my apartment building renovation business, and eventually I decided that I wanted a life like Steven Spielberg. I wanted to be just like him. How could I become a filmmaker? I used logic to figure out that the smartest thing to do would be to start with video. So I went out and bought a very expensive, broadcast, Sony camera ($65,000) and had a professional editing studio built into my home. Then I created a company called *VideoNet.* At the time the internet was a small idea and I liked the way 'VideoNet' kind of sounded like 'internet'. Who knew how web would grow at a lightning fast speed? I printed up some business cards and mailed them to the television stations on the east coast.

And one day the phone rang. It was ESPN regarding the Baltimore Oriole's player, Roberto Alomar's *'spitting incident'*, where a baseball player actually spit on an umpire. They needed a cameraman, with all the gear necessary to interview the head of the umpire's union. That was me! The Producer and I completed the interview, then we beamed it up to New York and the Producer of the show, Sal Paolantonio, said to me *"so who are you?"* I gave him my business card and the rest is history. From ESPN, I started getting work from ABC and CNN and then NBC, and then corporations like Disney Studios started to hire me for various projects. Many years later I achieved a certain level of success by knowing who the most talented people in the Philadelphia Metro area were and formed a reliable network of video professionals.

The beauty for our corporate clients is that the television broadcast world is much more technically demanding on the quality of the picture and sound. Focusing on this high level of professional standards is how I learned the business. The corporate world doesn't care so much about that. If it's in focus and if you can hear the people, they're happy. So I brought my broadcast experience and all this high quality, very expensive broadcast equipment into the corporate world, and my clients seem to love us for that.

I'm a filmmaker at heart, and yet I have the business acumen of profitably running my own companies for over thirty years. I can relate to the CEO, the Operations Manager, or the Marketing Director who has a budget they have to adhere to. They need to believe that this new tool (video) that I'm providing them is actually going to work and generate revenue for their company. That's not a trust that I take for granted. I take that very seriously, and appreciate the confidence they have in me and my team of professionals. That's my story and I'm sticking to it.

CONCLUSION

So...video makes sense because it changes the way people think. And how they think directs how they buy. If you want to become more profitable as a company, and who doesn't, then you need to figure out how you want to bring video into your operation.

Start out walking, then when you see it working you can begin to sprint. When you decide where to bring it in seek out a Producer who can provide the necessary technical advice and story-telling capabilities to make it work for years to come. But remember, money doesn't buy happiness. It has been said that *true happiness* comes through the fidelity of a worthy purpose. Go out and grow a fanatical base of devoted customers. That's qualifies as a worthy purpose. In my business world, people fit into one of four categories: Suspects, Prospects, Clispects, or Loyalists.

A *Suspect* is anybody with a pulse who knows about VideoNet, and may or may not necessarily be our target market. A *Prospect* is someone who we've made contact with and determined that they possibly are our target market and may need our services some day in the future. *Clispects* (yes, I made that word up) are people who have done business with us, but then forget about us six months later, and would clearly use someone else if they come along at the right time. A *Loyalist* is what we want. That's a client who has done business with us, who knows us, who trusts and appreciates us, who believes we delivered superior service, and who always refers business to us and will never consider using anyone else. That's what we are out to attract. That's the kind of business you can build too. Let me know if there is anyway in which we can help.

THANKING EVERYONE WHO MADE THIS BOOK POSSIBLE

I will start with my wife Maria, and my two daughters Brooke and Chelsea, mom and dad.

People who have influenced my decision to put into writing what I have in my head would include: Marc LeBlanc, Jason Miller, Mike Lieberman, Eric Kieles, Bruce Mowday and Ellen Langes Campbell.

People who have highly influenced my thinking regarding many of the marketing issues discussed in this book: Kirk Hanes, George Brodt, Denny Gernert, Brent Gernert, Mark Manjardi, Lou D'Angeli, Barry Burkhardt, Chuck and Simone Politi, Debbie Suedmeier, Gino Crognale, Tom Corgnale, Dick Flaville, Beau Westhoff, Russ Gwynne, Richard Stellar, Scott Beavers, Tim Boyle.

People who have been instrumental in our success as a company over the years: Ric Edevane, Robert Cooper, Stanley Lieberman, Anne Bachrach, Kerry Cargill, Ian Alexander, Ed Buffman, Cameron Shipman, David Newman, Alan Pratt, Matthew

Lauyer, Scott Messer, Ed Stevens, Chris Ambolino, Mike Boornazian, Matt Radico, Steve Rhykerd, Steven Bayles, Joe Hennigan.

Loyalists (see page 93 for description), who are the engine of our corporate success would include, at the time of printing this book: Ray Hollinger, Bryan Gernert, Anne Marie Heil, Becky Austill, Alistair Linton, Eugene Andreassi, Guy Gosselin, Jammie O'Brien, Barbara Lehman, Tore Fiore, Tim Connor, Gary Smith, Diane Gring, Neil Beresin, Marc & Devon Heppe, Tony Girifalco, Art Dodge, Bob Delaney, Marty Ferrill, Jim O'Brien, Roy De La Mar, Michael Letts, Cheryl Wade, Maurice and Ann Glavin, Ted Van Beuren, Mary McLoughlin, Dino Quaglietta, Emily Baker, Sherri Velte, Cricket Capucci, Valerie Nolan, Daz Connell, Elizabeth Florer, Christina Gurtman, Susan Toombs, Joan Fiddle-Fereder and others.

Made in the USA
Monee, IL
04 November 2026

67380522R00069